U.S. Department
of Transportation

**Federal Aviation
Administration**

A First Look at Salience and Distinctiveness of Fly-Over and Fly-By Waypoint Symbology

Project Memorandum No. DOT-VNTSC-FAA-00-23
September 2000

Divya C. Chandra
Daniel J. Hannon

OPERATOR PERFORMANCE AND SAFETY ANALYSIS DIVISION

Research and Special Programs Administration
John A. Volpe National Transportation Systems Center
Cambridge, MA 02142-1093

Approved for Distribution:

U.S. Department of Transportation
Research and Special Programs
Administration
Transportation Systems Center
Cambridge, MA 02142-1093

This document contains preliminary information subject to change. It is considered an internal working paper with a selected distribution limited to project participants and appropriate supervisory staff. It is not a formal referable report.

Preface

This report presents the findings of a short-term empirical study on human factors issues related to the selection of symbology for fly-over and fly-by waypoints on aeronautical charts. At issue were the international standards for the depiction of these symbols. The symbols in use by the United States (US) and the International Civil Aviation Organization (ICAO) are currently in conflict. Results of this study were required for the ICAO Obstacle Clearance Panel (OCP), which planned to resolve this conflict at a June 2000 meeting in Madrid. Because fly-over waypoints are often used to ensure obstruction clearance, it is critical for safety of flight that the symbols are unambiguous and salient to pilots.

This work was conducted at the Volpe National Transportation Systems Center (Volpe Center) under the sponsorship of the Federal Aviation Administration's (FAA) Office of the Chief Scientific and Technical Advisor for Human Factors. The research was performed at the request of the FAA Office of Flight Standards. Tom McCloy served as the FAA program manager. The authors consulted with Lynn Boniface (FAA Flight Standards), Colleen Donovan (FAA Aircraft Certification), Kathy Abbott (FAA Certification and Regulation), Jim Gregory (ICAO), Jim Terpstra and others at Jeppesen, Inc., and John Moore, Eric Secretan, and others at National Oceanic and Atmospheric Administration (NOAA) to complete the work. The authors would especially like to thank NOAA for constructing the charts used in the study.

The document was prepared by the Operator Performance and Safety Analysis Division of the Office of Research and Analysis at the Volpe Center. It was completed under the Division's Flight Deck Technology Human Factors program.

Table of Contents

Preface i

Table of Contents ii

List of Figures iii

List of Tables iv

Executive Summary v

Introduction 1

Research Issues and Focus 1

Method 3

Results 13

Summary and Discussion 21

Conclusion 21

Distribution List 22

Appendix A: List of Mock Charts and Procedures 23

Appendix B: Excerpts from Surveys 24

List of Figures

Figure 1. Sample mock chart using the US NOAA fly-over and fly-by waypoint symbols. 6

Figure 2. Sample mock chart using the ICAO fly-over and fly-by waypoint symbols. 8

Figure 3. Sample mock chart using the IATA Compromise fly-over and fly-by waypoint symbols. 10

Figure 4. Sample legend and test symbol from the ICAO-style symbol survey. 12

Figure 5. Pilot confidence in rating ICAO-style compulsory/on-request symbols. 15

Figure 6. Pilot confidence in rating NOAA-style compulsory/on-request symbols. 16

Figure 7. Pilot confidence in rating Jeppesen-style compulsory/on-request symbols. 17

Figure 8. Pilot confidence in rating ICAO-style fly-over/fly-by waypoint symbols. 18

Figure 9. Pilot confidence in rating NOAA-style fly-over/fly-by waypoint symbols. 19

Figure 10. Pilot confidence in rating Jeppesen-style fly-over/fly-by waypoint symbols. 20

List of Tables

Table 1. Symbols in use for fly-over and fly-by waypoints by the US (NOAA) and ICAO................1

Table 2. Research issues for fly-over and fly-by waypoint symbols. 2

Table 3. Test symbol sets for mock-chart search task..4

Table 4. Illustrations of the rules tested in the survey task. 5

Table 5. Accuracy of finding symbols in the mock chart task for each test condition. 13

Table 6. Outcome of OCP discussion on fly-over and fly-by waypoint symbols 21

Executive Summary

The purpose of this project was to conduct an empirical evaluation of a set of symbols that could represent "fly-by" and "fly-over" waypoints on aeronautical charts. At issue were the international standards for the depiction of these symbols. The symbols in use by the United States (US) and the International Civil Aviation Organization (ICAO) are currently in conflict. The Federal Aviation Administration (FAA) Office of Flight Standards requested that the Volpe Center conduct a short-term (2-month) empirical study to evaluate the human factors issues associated with both US and ICAO fly-over and fly-by waypoint symbols. The data were required in advance of the June 2000 meeting of the ICAO Obstacle Clearance Panel (OCP) in Madrid, at which the conflict between the two symbol sets was to be resolved.

A fly-by waypoint is one where the pilot is required to use turn anticipation to avoid overshoot of the next flight segment. A fly-over waypoint precludes any turn until the waypoint is overflown, and is followed by an intercept maneuver of the next flight segment. Fly-over waypoints are often used to ensure obstruction clearance, so it is critical for safety of flight that the symbols used to depict these waypoints are unambiguous and salient to pilots.

This study focused on two of the issues related to the selection of these symbols: the salience of these symbols in a cluttered chart context and the distinctiveness of the features in fly-over and fly-by symbols. It was a "first look" study in that it did not evaluate these issues in depth and it did not directly examine other important issues at all (e.g., electronic display of the symbols).

The study was conducted with a small sample of pilots from US airlines, some of whom had international flight experience. The pilots performed two tasks. Their first task was to search for fly-over and fly-by waypoints on a small number of mock charts. The mock charts were based on the US chart format. In particular, waypoints at the end of a runway were smaller than waypoints drawn elsewhere on the chart. Three sets of symbols were tested: the standard ICAO symbols, the standard US symbols, and a compromise symbol set which used the US fly-over symbol and the ICAO fly-by symbol. Jeppesen fly-over and fly-by waypoint symbols were not tested in the mock-chart task.

Results from the mock chart task supported the compromise symbol set. They indicated that, for fly-over waypoints, detection accuracy is significantly reduced for small ICAO fly-over waypoint symbols, as compared with similar small symbols that have surrounding circles, which are used by the US. Performance with the larger fly-over waypoint symbols did not differ significantly across the three tested symbol sets. For fly-by waypoints, detection accuracy was best with the ICAO fly-by symbols. Note, however, that there were no small fly-by waypoints in any of the mock charts.

The second task for the pilots was to complete surveys in which they were shown fictitious symbols that were created from the same rules and features that are used to draw either the US, ICAO, or Jeppesen fly-by and fly-over waypoint symbols. Pilots rated their confidence in interpreting the meaning of these fictitious symbols based on the legend for that particular set of symbols. Results from the survey task showed that pilots could reliably discern the intended meaning of the fictitious symbols based on each of the legends, but they were not inclined to generalize the symbol-feature rules broadly.

The results of this study were presented to the ICAO OCP at the June 2000 meeting. This group supported the addition of a circle to the ICAO fly-over symbol based on this research.

Introduction

The purpose of this project was to conduct an empirical evaluation of a set of symbols that could represent "fly-by" and "fly-over" waypoints on aeronautical charts. A fly-by waypoint is one where the pilot is required to use turn anticipation to avoid overshoot of the next flight segment. A fly-over waypoint precludes any turn until the waypoint is overflown, and is followed by an intercept maneuver of the next flight segment.[1] Fly-over waypoints are often used to ensure obstruction clearance, so it is critical for safety of flight that the symbols are unambiguous and salient to pilots.

In the United States (US), government aeronautical charts are produced by the National Oceanic and Atmospheric Administration (NOAA). By international agreement, charts produced by member states around the world follow the standards set by the International Civil Aviation Organization (ICAO). In the case of symbols for fly-over and fly-by waypoints, NOAA and ICAO symbols are in conflict, as shown in Table 1.[2] In particular, the ICAO fly-over symbol is nearly identical to the US fly-by symbol already in use. This is unacceptable because pilots could easily become confused if they see such similar symbols representing two different flight paths.

The Federal Aviation Administration (FAA) Office of Flight Standards requested that the Volpe Center conduct a rapid, short-term (2-month) empirical study to evaluate the human factors issues associated with both US and ICAO fly-over and fly-by waypoint symbols. The data were required in advance of the June 2000 meeting of the ICAO Obstacle Clearance Panel (OCP) in Madrid, at which the conflict between the two symbol sets was to be resolved.

Jeppesen Sanderson Inc. (Jeppesen), a private company based in the US that supplies aeronautical charts to the majority of the world's pilots has its own symbols for fly-over and fly-by waypoints as well. The Jeppesen symbology is similar to the US NOAA symbology in that the fly-over waypoint has an exterior circle, while the fly-by symbol does not. However, the shape of the Jeppesen four-pointed star is different. Because international standards were the primary issue, this research did not focus on evaluation of the Jeppesen symbols.[3]

Research Issues and Focus

The first task was to identify and clarify issues that affect the choice of fly-over and fly-by waypoint symbols. These issues are listed below in Table 2 (next page). Many of the issues are inter-related. For example, size of the symbol may affect how salient the symbol is on a cluttered chart.

This study focused on two issues: the salience of these symbols in cluttered chart context and the distinctiveness of fly-over and fly-by symbols. Symbols that are *salient* in a cluttered context are easy to find. It is important that fly-over symbols in particular be easy to find (i.e., salient) because they may denote an obstruction to safe flight. Symbols that are *distinctive* are easy to separate from other symbols, which may have only a single discriminating feature. (For example, in the US symbology, a fly-over waypoint symbol is constructed by circumscribing the symbol for a fly-by waypoint with a circle.) Pilots

Table 1. Symbols in use for fly-over and fly-by waypoints by the US (NOAA) and ICAO.

	US Symbols (NOAA)	*ICAO Symbols*
Fly-Over Waypoint	⊕ (star in circle)	◆ (star)
Fly-By Waypoint	✦ (star)	◇ (outline star)

Table 2. Research issues for fly-over and fly-by waypoint symbols.

Distinctiveness	Are the fly-over/fly-by symbols easily distinguished from each other outside the context of a chart? Does symbol size affect how distinctive the symbols are? Do feature-based rules (e.g., one is circled, the other is not) reinforce the meaning of the symbols, or are the symbols seen as unique entities specified in the legend, with no general relationship to one another?
Salience in Context	Are the symbols easy to find within a cluttered background such as that of a real approach plate? How salient are the symbols in a chart used for operations under visual flight regulations (VFR)? How does the type of clutter affect the salience of the symbol? Does the complexity of the depicted procedure affect the salience of the symbol?
Influence of Pilot Experience and Training	Does pilot experience/training affect the types of symbols that are easiest for them to locate and use? For example, do the same symbols work best for the general aviation pilot and the commercial pilot? Do pilots who always use a particular type of chart have trouble understanding symbols based on unfamiliar conventions?
Electronic Display Limitations	Fly-over/fly-by waypoints will be depicted on electronic map displays. Some of these displays are very small (e.g., comparable to the 3- inch unit used for storm scope displays). Many of these displays have very low resolution. Are some symbols inappropriate for electronic display because the features will be difficult or impossible to discriminate on low-resolution or very small displays?
Symbol Size Limitations	Large symbols are easier to see than very small symbols, but symbols must not be too large on aeronautical charts because of the additional space required and the resulting chart clutter. How small can the symbols be before their utility is severely reduced? Are some symbol designs easier to use in small formats than others? How does display resolution affect the minimum useful symbol size?
Compatibility with Other Symbols	a) **Moving Map Display Symbology:** Pilots who fly FMS-equipped aircraft may look to a paper chart for confirmation that the FMS is flying over or by a waypoint correctly. Therefore, symbols may need to be depicted so that they are readily usable in either paper or electronic format. Symbols for the two formats must also be consistent with each other. Are the moving map display symbols similar to the paper version of the symbols to eliminate any potential confusion? b) **Symbol Combinations:** In some cases, fly-over/fly-by symbols will be co-located with other types of markers on the charts (e.g., ground-based navigation aids, and/or reporting points). In some chart formats today, there are strategies for combining these symbols. For example, the symbol shape may represent the type of navigation aid, and its fill color (black or white) may represent whether that navigation aid is a compulsory reporting point or not. Is there a well thought-out approach as to how to display co-located waypoints/navigation aids?

must be able to determine whether a waypoint is fly-over or fly-by so that the appropriate flight path is selected. These two issues were chosen both because they were fundamental, and because the authors felt they could be addressed to some extent with a short-term study.[4]

The study also touched upon two other issues mentioned in Table 2 although it was not specifically designed to address these issues: (a) the compatibility of the fly-over and fly-by waypoint symbols with other symbols and (b) symbol size limitations.

The compatibility of the fly-over and fly-by waypoints with only one other symbol, a triangle that denotes whether the location is a compulsory reporting point or not, was examined. The significance of a compulsory reporting point is that pilots are required to contact Air Traffic Control (ATC) when they are not in radar coverage and they pass a compulsory reporting point. Non-compulsory, or "on-request" reporting points are named locations that pilots may have to report passing if requested by (ATC). Both the US and ICAO denote compulsory reporting points with filled (black) triangles and on-request reporting points with unfilled triangles. In some cases, compulsory or on-request reporting points may be co-located with a fly-over or fly-by waypoint.

The study addressed the symbol size issue in a limited way as well. This was not a planned comparison, but it arose because of the way in which the mock charts were constructed. The mock charts were constructed by NOAA, which uses two symbol sizes in its charts. The standard fly-over or fly-by symbol is approximately 5 mm in diameter in the plan view of an actual chart. However, if any waypoint symbol is drawn at the runway, the symbol size is reduced to approximately 3 mm so that the runway is visible.

Method

A two-part paper-and-pencil study was conducted with a small set of airline pilots from the US, some of whom had international flight experience, to address the issues mentioned above. The first part of the study assessed symbol salience by examining pilots' ability to find the US and ICAO fly-by and fly-over symbols on mock aeronautical charts and procedures. The mock charts were constructed by NOAA and based on their format (see Figures 1, 2, and 3 for examples). The second part of the study looked at the distinctiveness between the fly-over and fly-by symbols. Pilots completed surveys in which they were shown fictitious symbols that were created from the same rules and features that are used to draw either the US, ICAO, or Jeppesen fly-by and fly-over waypoint symbols. Pilots rated their confidence in interpreting the meaning of these fictitious symbols based on the legend for that particular set of symbols. The goal of this task was to test whether the symbol-feature rules (e.g., an outer circle indicates a "fly-over" waypoint) are evident without instruction and applied with confidence.

Mock-Chart Search Task

Objective

The goal of this task was to assess pilot speed and accuracy in finding fly-over and fly-by waypoints on representative mock aeronautical charts and procedures. Due to scheduling constraints, the task had to be performed within a 30-minute time limit, so pilots had to work fairly quickly, and the accuracy of finding the symbols varied.

Subjects

Eleven airline pilots from a US airline participated. Their flight experience ranged from 3,800 to 16,000 flight hours, with an average of 7,973 flight hours. Three of the pilots flew international routes, four flew aircraft that were capable of area navigation (RNAV), and three flew with moving map displays. All participants had normal or corrected-to-normal vision.

Table 3. Test symbol sets for mock-chart search task.

	US Symbols (NOAA)	ICAO	IATA Compromise
Fly-Over Waypoint	⊕	⊕	⊕
Fly-By Waypoint	✧	✧	✧

Materials

Eight instrument approach procedure (IAP) charts, one standard arrival route (STAR), and two standard instrument departures (SIDs) were used in the study. Appendix A contains a list of the procedures used in the study. Due to time constraints, informal expert opinion was solicited in order to obtain a representative selection of procedures with fly-over and/or fly-by waypoints. The procedures were selected by a team of aeronautical cartographic specialists at the NOAA and instrument-rated pilots at the Volpe Center.

Three versions of each chart were constructed for the study using each of the symbol sets shown in Table 3 above. The compromise in the third column of Table 3 was proposed by a representative to the ICAO OCP from the International Air Transport Association (IATA).

Figures 1, 2, and 3 are examples of the charts and procedures used in the study. The chart in Figure 1 is drawn using the US (NOAA) fly-over and fly-by waypoint symbols, the chart in Figure 2 shows use of the ICAO symbols, and the chart in Figure 3 shows use of the IATA compromise symbol set. Notice that only the fly-over and fly-by waypoint symbols change. The basic chart format is the one used by NOAA, which has both large and small waypoint symbols. The fly-over waypoints are typically located at the runway end and at holding patterns.

Procedure

Test booklets were constructed with four practice charts and three sections of test charts (i.e., one section for each of the three symbols sets shown in Table 3). Within a test section, all 11 selected procedures (i.e., instrument approaches, and standard departure or arrival routes) were drawn with the same fly-over and fly-by waypoint symbols. Each procedure filled a complete page. The order of the 11 procedures was randomized within the section. The order of test chart sections was also randomized between test booklets. The test booklet was arranged in a binder such that the procedure always appeared on the right-hand page. To the left of the procedure was a separate page with a symbol legend for that procedure. The legend showed six different symbols: (1) fly-over waypoint, (2) fly-by waypoint, (3) VOR/DME, (4) compulsory reporting point, (5) on-request reporting point, and (6) VORTAC.

For each procedure, the pilot used a highlighter pen to mark all occurrences of *one* of these six types of symbols.[5] The symbol of interest was identified on the left page, above the legend. For example, the pilot might be asked to highlight all fly-over waypoints in the chart on the right-hand page. Although the fly-over, fly-by waypoint symbol distinction was the primary concern of this study, the other symbols were included to encourage the subjects to review each question and the procedure carefully. Of the 11 cases in each section, pilots were asked to search four times for a fly-over waypoint, four times for a fly-by waypoint, twice for any of the other symbols, and once for fly-over waypoints when there were none on the chart.

Due to the limited time available for constructing the mock charts, the same 11 charts appeared in each test section, but with different symbols for the fly-over and fly-by waypoint, as appropriate. In order to minimize pilot familiarity with the charts, the specific symbol the pilot looked for on a given procedure was changed between the test sections. For example, if the pilot had to look for fly-by waypoints on the NOAA version of a particular chart, then the pilot would be asked to look for fly-over waypoints on the ICAO version.

Due to scheduling constraints for the study, pilots were given 30 minutes to complete the test booklet. They began by completing the four practice trials. At this point, they could ask questions if necessary. Then they proceeded through the test booklet (33 test trials) at their own pace. If they were able to complete these trials, they were asked to continue on to Part 2 (Survey Task), described below. The study was run in a classroom setting, but participants were not allowed to consult one another.

In actual flights, pilots would be able to review the charts as many times as they liked for as long as they liked. For the study, the amount of time the pilots spent reviewing a procedure was not limited, but they were asked not return to a procedure after they had highlighted the symbols on that procedure.

Survey Task

Objective

The goals of this task were to determine (a) whether pilots could extract symbol-feature rules from a legend without explicit instruction, and (b) how confidently they applied the rules to fictitious symbols. Four specific rules, which are used to draw existing symbols, were evaluated. These are listed and illustrated in Table 4 below.

Table 4. Illustrations of the rules tested in the survey task. Note: Jeppesen air space fix/waypoint symbology Copyright 2000 Jeppesen Sanderson, Inc.

Rule	Examples	
ICAO Fill Rule Fill/No-fill of the 4-points of the star distinguishes between fly-over and fly-by waypoints.	Fly-Over	
	Fly-By	
ICAO and NOAA Triangle Rule A filled/unfilled triangle distinguishes between compulsory and on-request waypoints.	Compulsory	and
	On-Request	and
Jeppesen and NOAA Circle Rule Presence/absence of a circle around symbol distinguishes between fly-over and fly-by waypoints.	Fly-Over	and
	Fly-By	and
Jeppesen Fill Rule A filled/unfilled shape distinguishes between compulsory and on-request waypoints.	Compulsory	and
	On-Request	and

Figure 1. Sample mock chart using the US NOAA fly-over and fly-by waypoint symbols (next page).

RNAV RWY 14

SALISBURY-OCEAN CITY WICOMICO REGIONAL (SBY)
SALISBURY, MARYLAND

NOS Waypoint Symbols

APP CRS	Rwy Idg	5500
139°	TDZE	49
	Apt Elev	52

⚠ NA

MISSED APPROACH: Climb to 500, then climbing left turn to 2500 via course 013° to EZIZI WP and hold.

PATUXENT APP CON ★	SALISBURY TOWER ★	GND CON	CLNC DEL	UNICOM
127.95 314.0	119.425 (CTAF)	123.775	118.55 (2230-0600) 123.775 (0600-2230)	122.95

NOT FOR NAVIGATION

ELEV 52

CATEGORY	A	B	C	D
GLS DA	NA			
LNAV/VNAV DA	NA			
LNAV	400-1 351 (400-1)			400-1¼ 351 (400-1¼)
CIRCLING	460-1 408 (500-1)	520-1 468 (500-1)	520-1½ 468 (500-1½)	620-2 568 (600-2)

REIL Rwys 5, 14 and 23
MIRL Rwy 5-23
HIRL Rwy 14-32

Orig 00111
AL-977 (FAA)

38°20'N-75°31'W

SALISBURY, MARYLAND
SALISBURY-OCEAN CITY WICOMICO REGIONAL (SBY)
RNAV RWY 14

Figure 2. Sample mock chart using the ICAO fly-over and fly-by waypoint symbols (next page).

RNAV RWY 14

SALISBURY-OCEAN CITY WICOMICO REGIONAL (SBY)
SALISBURY, MARYLAND

APP CRS 139°	Rwy Idg 5500 TDZE 49 Apt Elev 52		
⚠ NA	MISSED APPROACH: Climb to 500, then climbing left turn to 2500 via course 013° to EZIZI WP and hold.		

| PATUXENT APP CON ★
127.95 314.0 | SALISBURY TOWER ★
119.425 (CTAF) Ⓛ | GND CON
123.775 | CLNC DEL
118.55   123.775
(2230-0600) (0600-2230) | UNICOM
122.95 |

ICAO Waypoint Symbols

NOT FOR NAVIGATION

CATEGORY	A	B	C	D
GLS DA	NA			
LNAV/VNAV DA	NA			
LNAV	400-1 351 (400-1)			400-1¼ 351 (400-1¼)
CIRCLING	460-1 408 (500-1)	520-1 468 (500-1)	520-1½ 468 (500-1½)	620-2 568 (600-2)

REIL Rwys 5, 14 and 23 Ⓛ
MIRL Rwy 5-23 Ⓛ
HIRL Rwy 14-32 Ⓛ

Orig 00111
AL-977 (FAA)

38°20'N-75°31'W

SALISBURY, MARYLAND
SALISBURY-OCEAN CITY WICOMICO REGIONAL (SBY)

RNAV RWY 14

Figure 3. Sample mock chart using the IATA Compromise fly-over and fly-by waypoint symbols (next page).

RNAV RWY 14
SALISBURY-OCEAN CITY WICOMICO REGIONAL (SBY)
SALISBURY, MARYLAND

APP CRS	Rwy Idg	5500
139°	TDZE	49
	Apt Elev	52

⚠ NA

MISSED APPROACH: Climb to 500, then climbing left turn to 2500 via course 013° to EZIZI WP and hold.

PATUXENT APP CON ★	SALISBURY TOWER ★	GND CON	CLNC DEL		UNICOM
127.95 314.0	119.425 (CTAF)	123.775	118.55 (2230-0600)	123.775 (0600-2230)	122.95

NOT FOR NAVIGATION

CATEGORY	A	B	C	D
GLS DA		NA		
LNAV/VNAV DA		NA		
LNAV		400-1 351 (400-1)		400-1¼ 351 (400-1¼)
CIRCLING	460-1 408 (500-1)	520-1 468 (500-1)	520-1½ 468 (500-1½)	620-2 568 (600-2)

REIL Rwys 5, 14 and 23
MIRL Rwy 5-23
HIRL Rwy 14-32

Orig 00111
AL-977 (FAA)

38°20'N-75°31'W

SALISBURY, MARYLAND
SALISBURY-OCEAN CITY WICOMICO REGIONAL (SBY)
RNAV RWY 14

Subjects

Pilots from three US airlines participated. Some, but not all, participants had completed the mock chart task. A total of 13 pilots completed the NOAA- and ICAO-symbology surveys and 17 pilots completed the Jeppesen-symbology survey. The pilots who completed the Jeppesen-symbology surveys had on average 9,871 flight hours and the pilots who completed the NOAA and ICAO surveys had on average 8,823 flight hours. Of the pilots completing the NOAA- and ICAO-symbology surveys, four flew international routes, six flew aircraft that were capable of RNAV, and four flew with moving map displays. Of the pilots completing the Jeppesen-symbology survey, nine flew international routes, ten flew aircraft that were capable of RNAV, and ten flew with moving map displays. Participants reported their familiarity with different chart formats along a scale from 1(very unfamiliar) to 7 (very familiar). Nearly all reported that they were familiar or highly familiar with Jeppesen charts (average rating of 6.0) and most were somewhat familiar with NOAA charts (average rating of 4.2).

Materials

Three surveys were constructed, one using each of the symbol sets from NOAA, ICAO, and Jeppesen. Each survey showed the symbol legend, and then a series of test symbols in isolation. The NOAA and ICAO surveys each had 14 test symbols, and the Jeppesen survey had 12 test symbols. Pilots responded to two questions for each symbol, one on whether they thought the symbol represented a fly-over or fly-by waypoint, and the other on whether they thought the symbol represented a compulsory, or on-request waypoint. There was no flight context in the survey.

Some of the symbols on the surveys were taken directly from the appropriate legend. Other symbols were *realistic* combinations of the reporting point and waypoint symbology that did not explicitly appear in the legend. A few of the symbols were *fake*, in that they were not based on any of the legend symbols, nor were they associated with any other type of standard chart symbol. Fake symbols were included to evaluate how well the rules were generalized.

One sample legend and a question from an ICAO-symbol survey is shown in Figure 4. Appendix B contains the legends and test symbols from all the surveys. In Appendix B, each of the test symbols is labeled as to whether it is from the legend, realistic, or fake, and its intended meaning based on the given legend is specified.

Fly-Over Waypoint	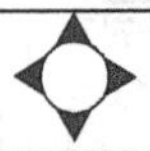	**Compulsory Reporting Point**	
Fly-By Waypoint	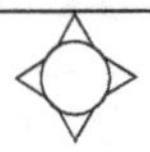	**On-Request (or "non-compulsory") Reporting Point**	

1 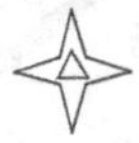	1 Definitely Fly-Over	2	3	4 Neutral (Can't Tell)	5	6	7 Definitely Fly-By
	1 Definitely Compulsory Reporting Point	2	3	4 Neutral (Can't Tell)	5	6	7 Definitely On-request Reporting Point

Figure 4. Sample legend and test symbol from the ICAO-style symbol survey. The question format was the same for all the surveys.

Procedure

Participants completed one or more of the surveys at their own pace. Their instructions were as follows:

Based on the legend below, please circle your responses to the following questions about "new" symbols. The legend is repeated on each page for reference. If you are unsure of a response, just mark one based on your first impression. There are no "correct" answers.

Results

Mock-Chart Search Task

The percent of symbols detected correctly for the non-waypoint symbols (e.g., VOR/DME), fly-by waypoints, and fly-over waypoints are shown in Table 5 below. Results for the fly-over waypoints are also broken down by the size of the symbol, large or small. The values in Table 5 were calculated by summing the total number of correctly highlighted symbols across all charts/procedures and dividing by the actual number of that type of symbol on all the charts/procedures.[6]

The values in Table 5 show that the best performance was found for the ICAO fly-by waypoints, where 100% were correctly detected. The poorest performance was found for the ICAO fly-over waypoints, where only 70% were correctly detected (across both large and small symbols).

Analyses of the results for fly-over and fly-by waypoints in Table 5 revealed statistically significant differences.[7] Specifically, there were significant differences in accuracy of detection between (a) the NOAA fly-over and ICAO fly-over waypoint symbols (86% versus 70%), (b) the IATA Compromise fly-over and ICAO fly-over waypoint symbols (86% versus 70%), and (c) the ICAO fly-by and ICAO fly-over waypoint symbols (100% versus 70%). This pattern of results indicates that pilots were more accurate at finding the NOAA and IATA-Compromise fly-over waypoints than they were at finding the ICAO fly-over waypoint. The results also indicate that pilots were significantly less accurate at finding the ICAO fly-over waypoints than the ICAO fly-by waypoints, a result not found with the NOAA and IATA-Compromise charts.

Further consideration of the results revealed that the less accurate performance with the ICAO fly-over waypoints was primarily due to the low accuracy of detecting small fly-over waypoint symbol (59%). In contrast, the small NOAA and IATA fly-over waypoints yielded much higher detection rates (84% and 80%, respectively). When the statistical analyses were re-computed to compare the fly-by waypoints

Table 5. Accuracy of finding symbols in the mock chart task for each test condition.

	NOAA	ICAO	IATA Compromise
Other symbols (VOR/DME, VORTAC, compulsory and non-compulsory reporting points)	88%	73%	95%
Fly-by Waypoints	90%	100%	93%
Fly-over Waypoints (all)	86%	70%	86%
Large Fly-over Waypoints	88%	85%	94%
Small Fly-over Waypoints	84%	59%	80%

against the small and large fly-over waypoints separately, significant differences were found only for the small fly-over waypoints. Specifically, there were statistically significant differences between the ICAO fly-by waypoints and the small ICAO fly-over symbols (100% versus 59%), between the small NOAA fly-over waypoints and small ICAO fly-over waypoints (84% versus 59%), and between the small IATA Compromise fly-over waypoints and small ICAO fly-over waypoints (80% versus 59%). No such statistically significant differences were found with the large fly-over waypoints.

Survey Task

Figures 5 through 10 show the results of the survey task. Figures 5, 6, and 7 show results for the question on whether the test symbol is thought to represent a compulsory or on-request reporting point. Figures 8, 9, and 10 show results for the question on whether the test symbol is thought to represent a fly-by or fly-over waypoint. The horizontal axis of each chart shows the scale on which pilots rated the symbols (from –3 to +3), where the extremes represent high confidence. Neutral scores are close to the center of the scale, near zero.

The average pilot rating for each test symbol is plotted as a horizontal bar on each chart. The length of the horizontal bar represents the magnitude of the average score, and the actual numerical average score is given at the end of the bar. The symbol that was tested is also drawn for reference at the end of each column. For example, in Figure 8, the average pilot rating for the test symbol shown in Figure 4 (a four-point star with an unfilled triangle in its center) is 0.8. This score implies that pilots, on average, felt that it represented a fly-by symbol based on the accompanying (ICAO) legend, but not very strongly. Scores that were not statistically significantly different from zero (at the probability level of 0.05 or better) are labeled "NS." Thus, the above mentioned score of 0.8 *was* statistically significant, even if pilot confidence was relatively low.

Within a given chart, the results for the test symbols are sorted vertically into two or three categories as appropriate. Some of the symbols are in the "legend" category, meaning that they were identified in the legend for that survey. Some of the symbols are in the "realistic" category, meaning that they were *realistic* combinations of the reporting point and waypoint symbology that did not explicitly appear in the legend. The last symbol category was "fake," meaning that they were not based on any of the legend symbols, and they were not associated with any other type of standard chart symbol. The symbol category is indicated by braces at the right side of the chart.

Note that in all cases where the results were statistically significant, the average pilot rating is in the "correct" direction, meaning that pilots could reliably discern the intended meaning of the fictitious symbol without explicit instruction. Symbols with the highest average confidence ratings are those based on the NOAA and ICAO triangle rule for denoting compulsory and on-request reporting points (see Figures 5 and 6).[8] The test symbols shown in Figures 5 and 6 are particularly interesting in that the confidence of ratings for the fake symbols was almost as high as the confidence of ratings for the real symbols. The Jeppesen fill rule for denoting compulsory and on-request reporting points was of moderate strength (see Figure 7).

The Jeppesen and NOAA circle rule for distinguishing between fly-over and fly-by waypoints was noticeable, but not strong (see Figures 9 and 10). Pilots were more confident applying this rule to realistic symbols than to fake ones, meaning that they did not generalize it. Also, pilots had more confidence in rating symbols with an exterior circle as "fly-over" than they had in rating symbols without an exterior circle as "fly-by." The ICAO fill rule for distinguishing between fly-over and fly-by waypoints was noticeable (Figure 8). This rule too was applied more confidently to realistic symbols than to fake symbols.

Figure 5. Pilot confidence in rating ICAO-style compulsory/on-request symbols.

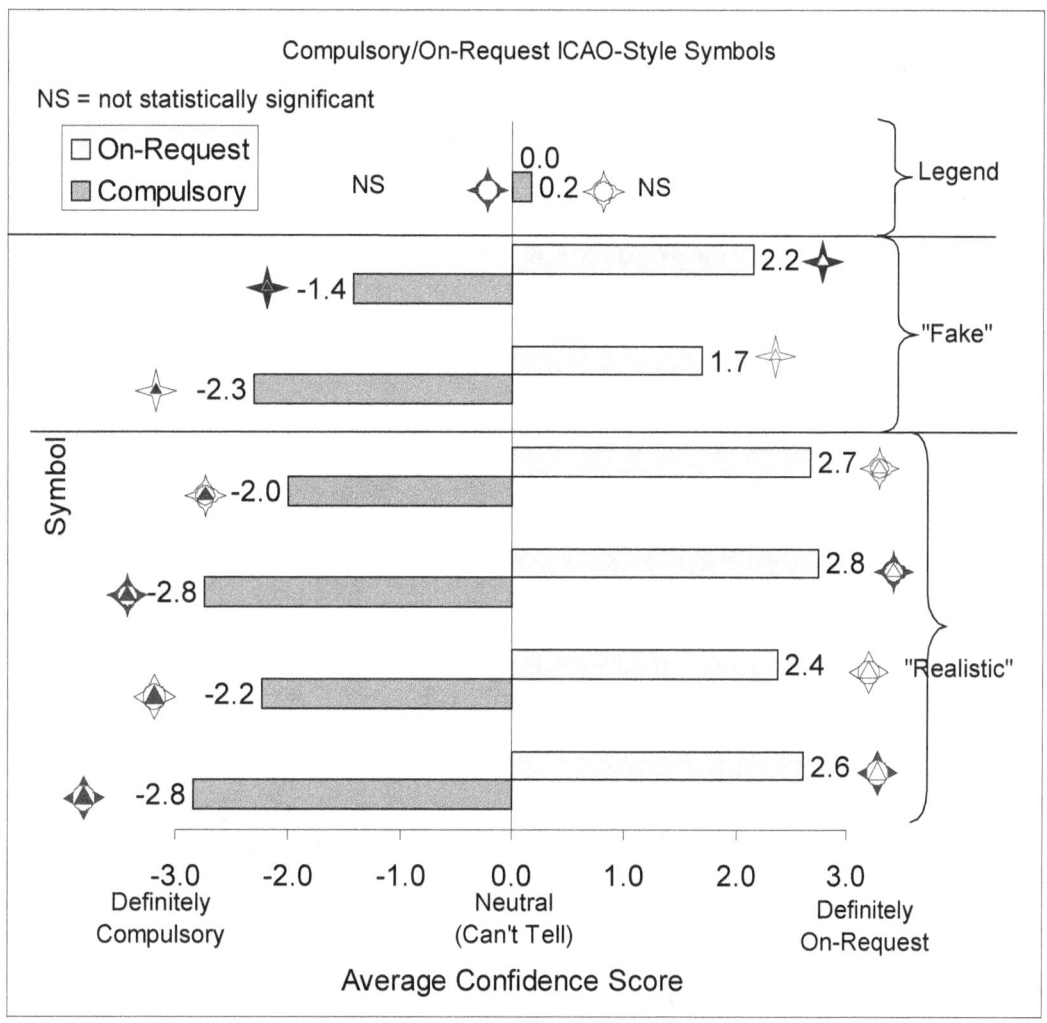

Figure 6. Pilot confidence in rating NOAA-style compulsory/on-request symbols.

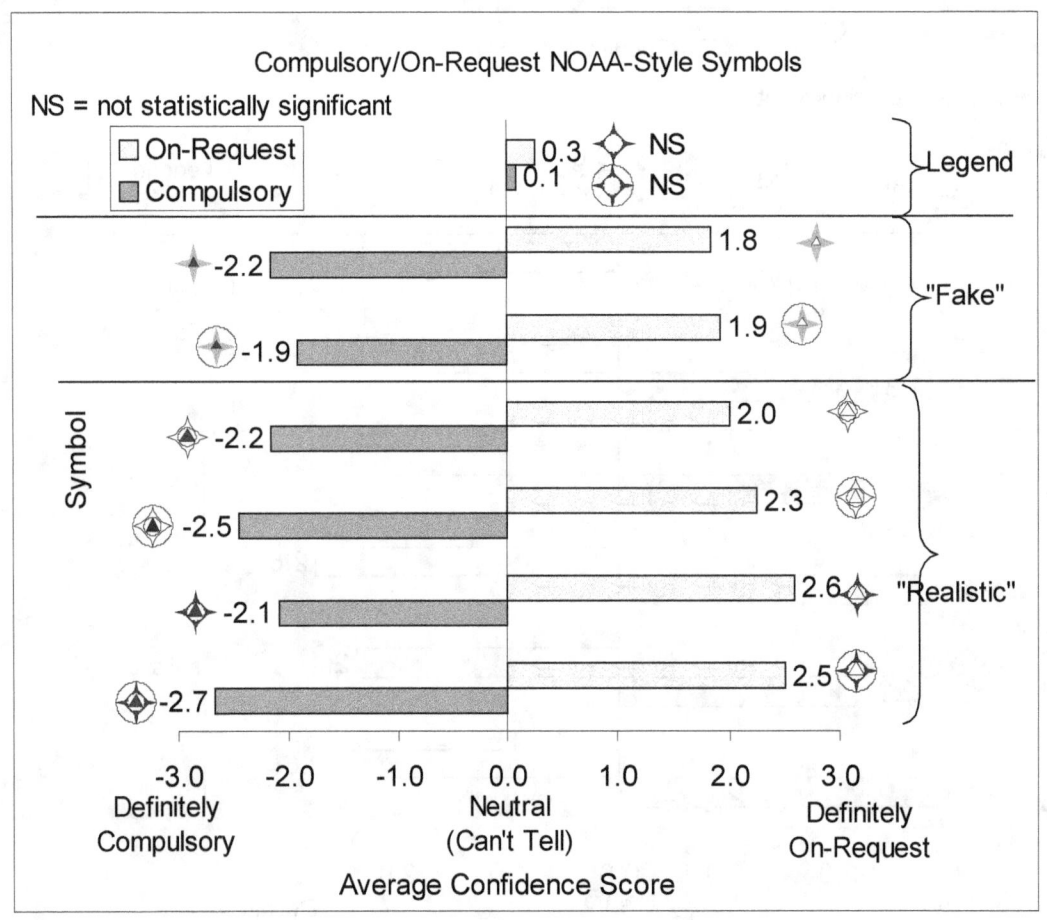

Figure 7. Pilot confidence in rating Jeppesen-style compulsory/on-request symbols. Note: Jeppesen air space fix/waypoint symbology Copyright 2000 Jeppesen Sanderson, Inc. Some symbols are reduced for illustrative purposes.

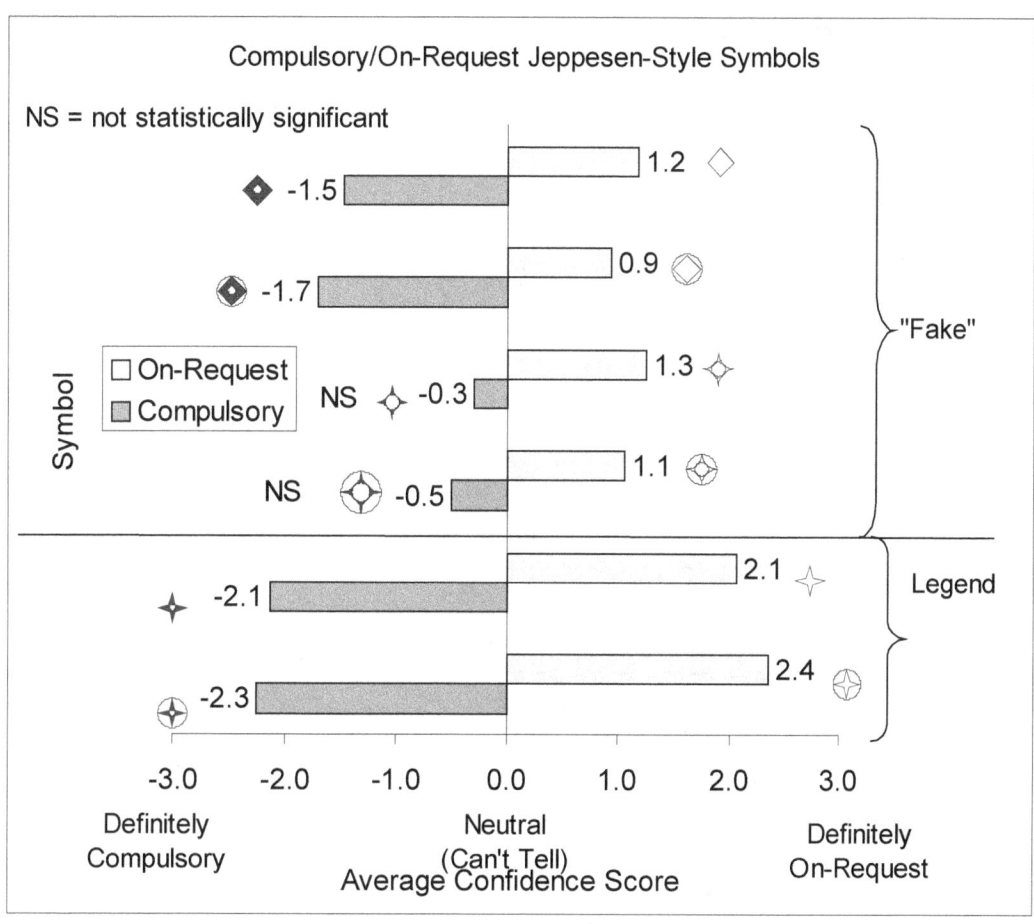

Figure 8. Pilot confidence in rating ICAO-style fly-over/fly-by waypoint symbols.

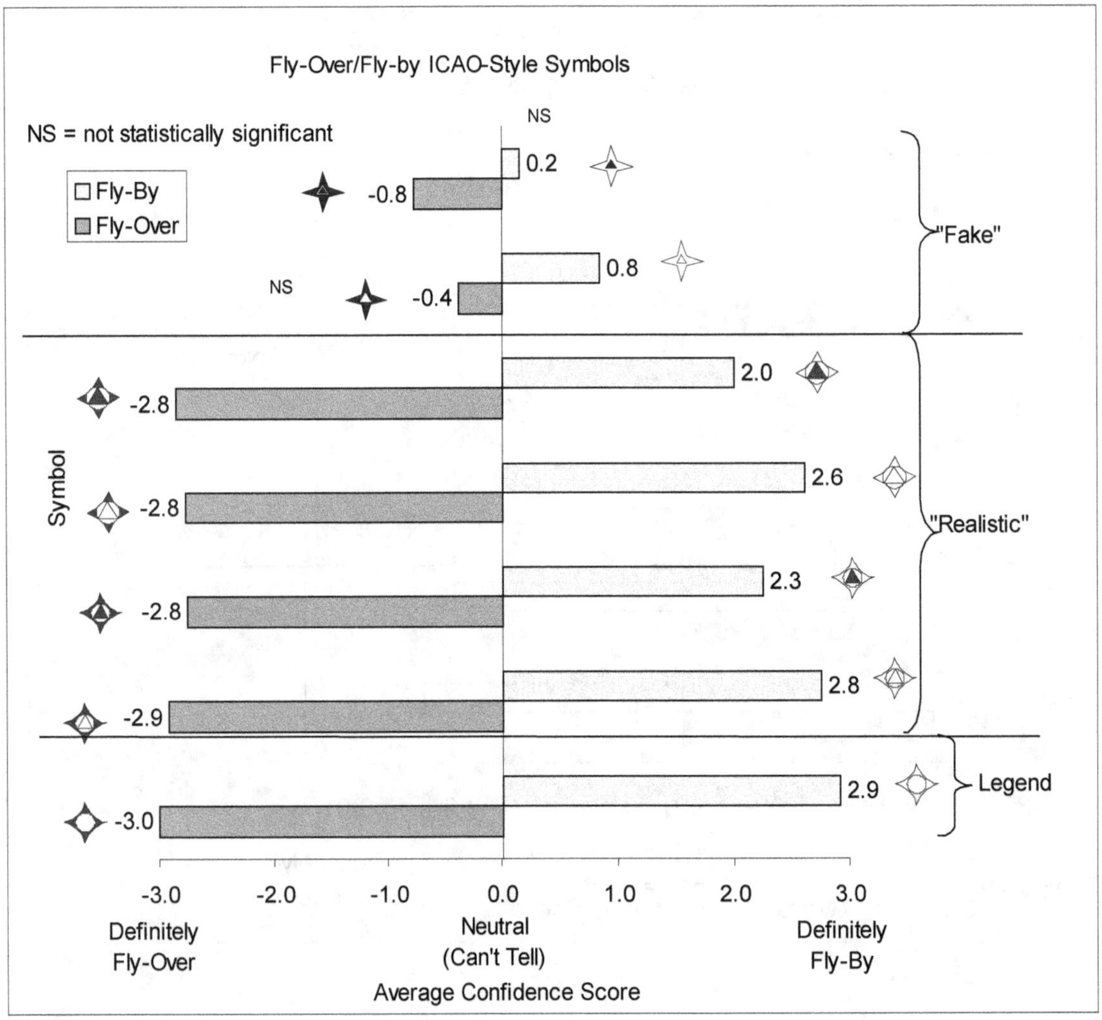

Figure 9. Pilot confidence in rating NOAA-style fly-over/fly-by waypoint symbols.

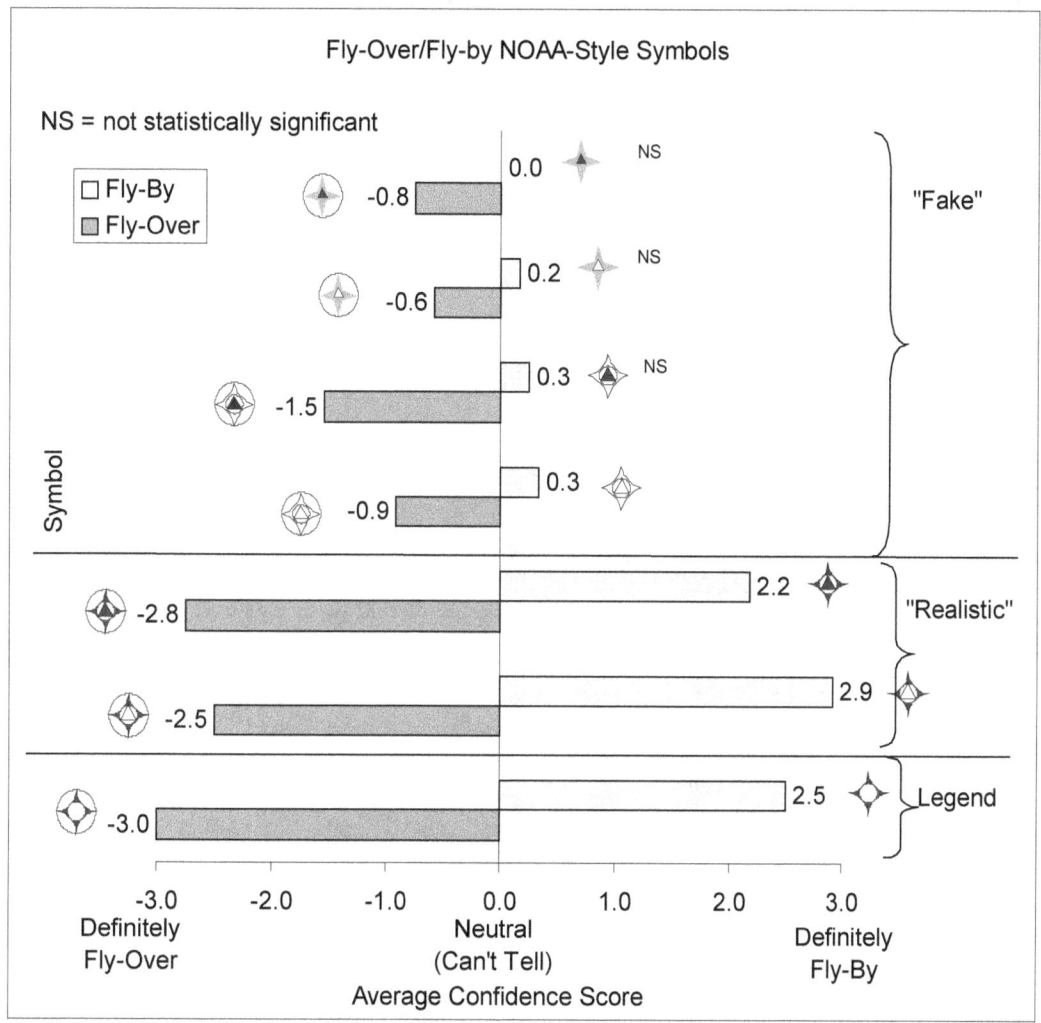

Figure 10. Pilot confidence in rating Jeppesen-style fly-over/fly-by waypoint symbols. Note: Jeppesen air space fix/waypoint symbology Copyright 2000 Jeppesen Sanderson, Inc. Some symbols are reduced for illustrative purposes.

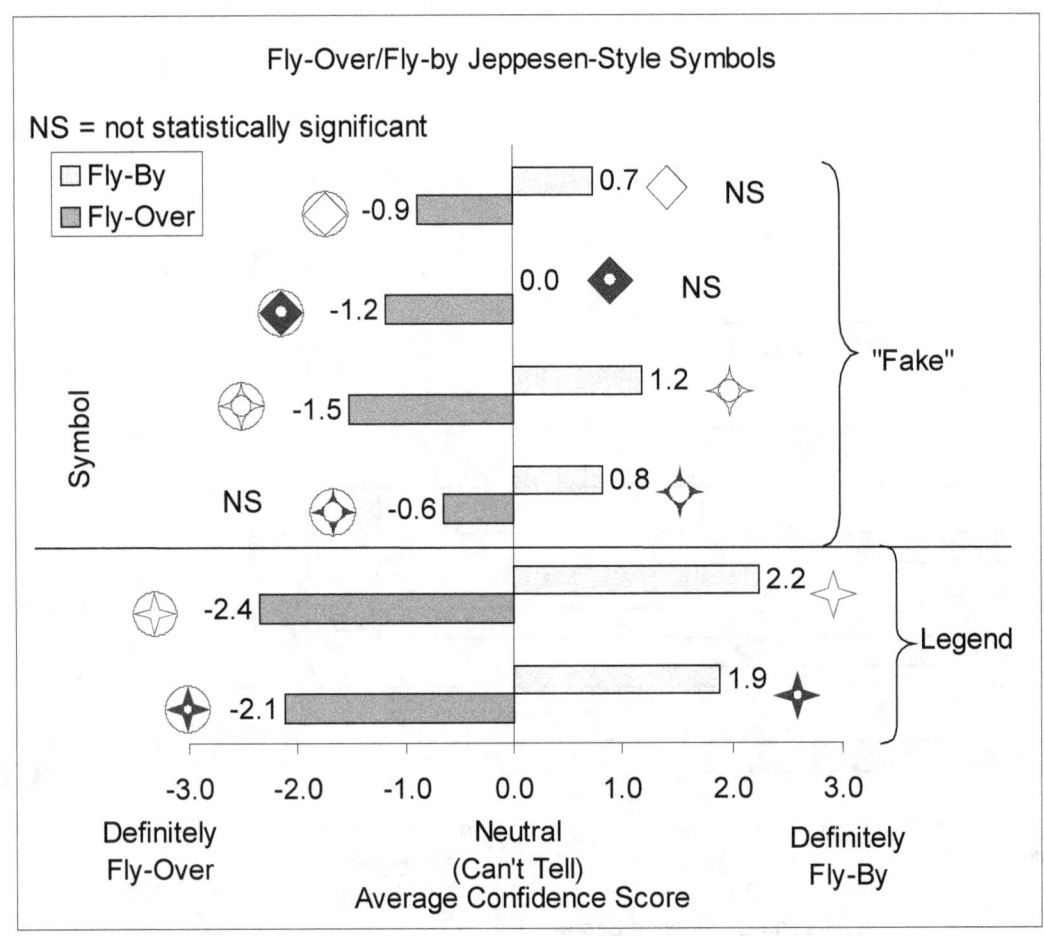

Summary and Discussion

Results from the mock chart task indicate that, for fly-over waypoints, detection accuracy is significantly reduced for small ICAO fly-over waypoint symbols, as compared with the NOAA and IATA compromise symbols, which both have surrounding circles. Performance with the larger fly-over waypoint symbols did not differ significantly across the three tested symbol sets. For fly-by waypoints, detection accuracy was best for the ICAO symbol. Overall, these results support the *IATA compromise* fly-over and fly-by waypoint symbols, which uses the ICAO fly-by waypoint symbol, and has a circle surrounding the fly-over waypoint symbol. Note, however, that there were no small fly-by waypoints in any of the mock charts. Also, note that the Jeppesen fly-over and fly-by waypoint symbols were not tested in the mock-chart task.

Results from the survey task showed that pilots could reliably discern the intended meaning of the fictitious symbols based on each of the legends, but they were not inclined to generalize the symbol-feature rules broadly. To many pilots, symbol-feature rules may not be important in learning and using chart symbols. The "triangle" rule, which denoted compulsory and on-request reporting points in the NOAA and ICAO conventions, was the strongest one. This may be because the triangle was seen as an independent object within a "joint" symbol rather than as a "feature" of a more complex symbol.

While the study provided useful input to the ICAO OCP, the research only addressed limited aspects of two important issues for the selection of aeronautical chart symbols. The study did not directly evaluate several other important issues at all (e.g., electronic display of symbols). Also, the data collection was conducted under less than ideal conditions, and only a limited number of charts/procedures were included in the test booklet. The data collection for the chart task, for example, should have ideally included a measure of how much time the pilot spent searching for symbols on each chart (i.e., a "response time"), but scheduling constraints dictated that data were collected under a time limit. Response time data could potentially reveal whether high accuracy also comes at the price of a longer time needed to search the chart. Finally, the number of participants in the study was small, and included only US airline pilots. A sample of pilots representing a broader range of experience, particularly a broader range of international flight experience would remove any potential bias in the results.

This preliminary evaluation could be improved by addressing the many constraints mentioned above. Also, the many research issues listed in Table 2 should be addressed in order to evaluate the selection of waypoint symbology in more depth.

Conclusion

Results of this study were presented to the ICAO OCP by Lynn Boniface (FAA) and Jim Terpstra (Jeppesen Sanderson) in June 2000. Based on these results, the OCP voted to add a circle to the ICAO fly-over symbol. The approved compromise symbols are shown along with the others tested in this study in Table 6 below.

Table 6. Outcome of OCP discussion on fly-over and fly-by waypoint symbols

	US Symbols (NOAA)	*ICAO Symbols*	*IATA Compromise*	*OCP Approved Compromise*
Fly-Over Waypoint	⊕	▲◆	⊕	⊕
Fly-By Waypoint	◆	◇	◇	◇

Distribution List

Tom McCloy, FAA AAR-100

Colleen Donovan, FAA AIR-130

Lynn Boniface, FAA AFS-420

Kathy Abbott, FAA AIR-105N

Jim Terpstra, Jeppesen Sanderson, Inc.

Jim Gregory, ICAO, Obstacle Clearance Panel

Erwin Lassooij, ICAO, Obstacle Clearance Panel

John Moore, NOAA

Eric Secretan, NOAA

Divya Chandra, Volpe Center

Dan Hannon, Volpe Center

Judith Burki-Cohen, Volpe Center

Donald Sussman, Volpe Center

Appendix A: List of Mock Charts and Procedures

The following US procedures were chosen by subject-matter experts for use in the mock-chart task portion of this study:

1. RNAV RWY 14, Ocean City, Maryland (OXB)
2. Playa One Departure (Pilot Nav), Houston, Texas (IAH)
3. Wylyy Three Departure (Pilot Nav) (FMS) (RNAV), Boston, Massachusetts (BOS)
4. RNAV RWY 10L, Pittsburgh Intl, Pennsylvania (PIT)
5. RNAV RWY 10, Pittsburgh/Allegheny County, Pennsylvania (AGC)
6. RNAV RWY 22, Georgetown, Delaware (GED)
7. Trudo One (FMS) Arrival, Milwaukee, Wisconsin
8. RNAV RWY 14, Salisbury, Maryland (SBY)
9. RNAV RWY 18, Morgantown, West Virginia (MGW)
10. RNAV RWY 4R, Boston, Massachusetts (BOS)
11. RNAV RWY 3, Charlottesville, Virginia (CHO)

Appendix B: Excerpts from Surveys

I. Excerpts from NOAA Survey

Legend

Fly-Over Waypoint	⊕	Compulsory Reporting Point	▲
Fly-By Waypoint	✦	On-Request Reporting Point	△

Test Symbol	Type Based on Legend	Interpretation Based on Legend	
(fly-over)	Legend	Fly-Over	Neither compulsory or on-request
(fly-over w/ triangle outline inside)	Fake	Fly-Over	On-Request
(fly-over w/ small triangle)	Fake	Fly-Over	On-Request
(fly-over w/ triangle outline)	Realistic	Fly-Over	On-Request
(fly-over w/ filled triangle inside)	Fake	Fly-Over	Compulsory
(fly-over w/ filled triangle)	Realistic	Fly-Over	Compulsory
(fly-over w/ small filled triangle)	Fake	Fly-Over	Compulsory
(fly-by)	Legend	Fly-By	Neither compulsory or on-request
(fly-by w/ triangle outline inside)	Fake	Fly-By	On-Request
(fly-by w/ small triangle)	Fake	Fly-By	On-Request
(fly-by w/ triangle outline)	Realistic	Fly-By	On-Request
(fly-by w/ filled triangle inside)	Fake	Fly-By	Compulsory
(fly-by w/ filled triangle)	Realistic	Fly-By	Compulsory
(fly-by w/ small filled triangle)	Fake	Fly-By	Compulsory

II. Excerpts from ICAO Survey

Legend

Fly-Over Waypoint	◇	Compulsory Reporting Point	▲
Fly-By Waypoint	◈	On-Request Reporting Point	△

Test Symbol	Type Based on Legend	Interpretation Based on Legend	
◇	Legend	Fly-Over	Neither compulsory or on-request
◇△	Realistic	Fly-Over	On-Request
◇△	Realistic	Fly-Over	On-Request
✦	Fake	Fly-Over	On-Request
◇▲	Realistic	Fly-Over	Compulsory
◇▲	Realistic	Fly-Over	Compulsory
✦	Fake	Fly-Over	Compulsory
◈	Legend	Fly-By	Neither compulsory or on-request
◈△	Realistic	Fly-By	On-Request
◈△	Realistic	Fly-By	On-Request
✦	Fake	Fly-By	On-Request
◈▲	Realistic	Fly-By	Compulsory
◈▲	Realistic	Fly-By	Compulsory
✦	Fake	Fly-By	Compulsory

III. Excerpts from Jeppesen Survey

Legend (Jeppesen air space fix/waypoint symbology Copyright 2000 Jeppesen Sanderson, Inc.)

Fly-By On-Request Waypoint	✧	Fly-By On-Request Fix	△
Fly-Over On-Request Waypoint	✧⃝	Fly-Over On-Request Fix	△⃝
Fly-By Compulsory Waypoint	✦	Fly-By Compulsory Fix	▲
Fly-Over Compulsory Waypoint	✦⃝	Fly-Over Compulsory Fix	▲⃝

Test Symbol	Type Based on Legend	Interpretation Based on Legend	
	Legend	Fly-Over	On-Request
	Fake	Fly-Over	On-Request
	Fake	Fly-Over	On-Request
	Legend	Fly-Over	Compulsory
	Fake	Fly-Over	Compulsory
	Fake	Fly-Over	Compulsory
	Legend	Fly-By	On-Request
	Fake	Fly-By	On-Request
	Fake	Fly-By	On-Request
	Legend	Fly-By	Compulsory
	Fake	Fly-By	Compulsory
	Fake	Fly-By	Compulsory

Notes

[1] The definitions of fly-over and fly-by waypoints were taken from the Aeronautical Information Manual used in the United States.

[2] The NOAA symbols were developed with reference to the Society of Automotive Engineers (SAE) Aviation Recommended Practices (ARP) No. 5289 *Electronic Aeronautical Symbols* October 1997.

[3] While the Jeppesen symbols were not the focus of the study, they were included in part of the study. Examples of the Jeppesen symbols can be seen in Appendix A.

[4] A comprehensive evaluation of these issues would require a study of interactions between issues listed in Table 2. For example, some symbols may be more salient for one type of chart clutter, but not another. To study the chart clutter issue fully would require a systematic variation of the background clutter. As another example, distinctiveness of the symbols may be affected by symbol size.

[5] For instrument approaches, pilots were asked to search only the *plan view* of accompanying test chart.

[6] On average, there were more fly-by waypoints (from 3 to 13) on a given chart than fly-over waypoints (1 or 2). Therefore, the percentages in Table 5 were not all calculated from the same denominator. Performance differences should be interpreted based on the statistical calculations, which take into account the number of data points for all cases, instead of the raw values in Table 5.

[7] Note that the 'Other' category of symbols was excluded from analysis since these symbols were not of interest for the study. All statistical tests were based on a probability level of 0.05 or lower, meaning that there is less than a 5% chance that a "significant" result was due to random factors.

[8] In both Figure 5 and Figure 6, the two test symbols that came from the legend are ambiguous with regard to whether they are "compulsory" or "on-request" waypoints because they do not have a triangle at all.

www.ingramcontent.com/pod-product-compliance
Lightning Source LLC
Chambersburg PA
CBHW081807170526
45167CB00008B/3368